6TH GRADE SCIENCE WORKBOOK
SPACE AND THE COSMOS

BABY PROFESSOR
EDUCATION KIDS

Speedy Publishing LLC
40 E. Main St. #1156
Newark, DE 19711
www.speedypublishing.com

Copyright 2018

Cosmos is a Greek word that describes an orderly and interconnected universe.

CROSSWORD PUZZLE

Use the clues to determine the appropriate term for each blank in this fun crossword puzzle.

Across

5. to turn around a center point.
7. a rocky space object of varying size.
8. the smallest planet in our solar system.
10. all the gases which surround a star or a planet.
11. a big ball of dirty ice and snow in outer space.
12. the path followed by an object in space as it goes around another object.

Down

1. the huge space which contains all of the matter and energy in existence.
2. the third closest planet to the Sun.
3. this planet has a moon called Triton.
4. having to do with the Sun.
6. a piece of stone or metal from space that falls to Earth's surface.
9. this planet is surrounded by over 1000 rings made of ice and dust.

Across

4. this planet tilts over so far on its axis that it rotates on its side.
5. an object that moves around a larger object.
6. once considered a planet, but demoted to a dwarf planet.
8. to move in an orbit or circle around something.
9. our home galaxy in the universe.
10. this planet is sometimes called the evening star and is the closest planet to Earth.

Down

1. iron in the soil gives this planet its red color.
2. a celestial body that revolves around a star.
3. this planet is so large that all of the other planets could fit inside it.
5. made up of all the planets that orbit our Sun.
7. first animal sent up to outer space.
9. Earth's only natural satellite.

Across

5. a term used to describe anything that does not originate on Earth.
7. bright patches that are visible on the Sun's surface, or photosphere.
9. is an imaginary line through the center of rotation of an object.
11. point in the orbit of a planet or other celestial body where it is farthest from the Sun.
12. a mutual physical force of nature that causes two bodies to attract each other.

Down

1. part of the Sun's atmosphere just above the surface.
2. total or partial blocking of one celestial body by another.
3. an extremely bright meteor, can be several times brighter than the full Moon.
4. a collapsed core of a massive star.
6. a large grouping of stars.
8. a strand of cool gas suspended over the photosphere by magnetic fields, which appears dark as seen against the disk of the Sun.
10. theory that suggests that the universe was formed from a single point in space during a cataclysmic explosion about 13.7 billion years ago.

Across

3. is a giant ball of hot gas that creates and emits its own radiation through nuclear fusion.
5. a measure of the tilt of a planet's orbital plane in relation to that of the Earth.
8. the outer part of the Sun's atmosphere.
9. the amount of light emitted by a star.
10. a cloud of dust and gas in space, usually illuminated by one or more stars.
11. a natural or artificial body in orbit around a planet.

Down

1. a large ring of icy, primitive objects beyond the orbit of Neptune.
2. an astronomical unit of measure equal to the distance light travels in a year, approximately 5.8 trillion miles.
3. is a cataclysmic explosion caused when a star exhausts its fuel and ends its life.
4. a grouping of stars that make an imaginary picture in the sky.
6. a small particle of rock or dust that burns away in the Earth's atmosphere.
7. energy radiated from an object in the form of waves or particles.

Across

1. a person who flies in space, whether as a crew member or passenger.
3. is the largest moon of Saturn.
5. the shape is the Milky Way.
7. 186,282.39 miles/sec
10. is the most distant object visible to the naked eye.
12. the name of the brightest comet in the solar system

Down

2. a system or device that controls a vehicle's flight at a preset course and altitude.
4. the first man in space.
6. evaporated gas and dust from the surface of a comet nucleus that can expand to more than the size of a planet.
8. the largest asteroid, spherical in shape, and the size of Texas.
9. one rotation of a planet on its axis.
11. the central body of a comet.

Across

4.   the environment in which celestial objects exist.
6.   the closest of the four large moons of Jupiter.
7.   the icy moon of Pluto, about half the diameter of Pluto.
8.   a lunar or planetary phase wherein less than
     half the surface is illuminated.
9.   beyond the sun.
12. minute particles floating in space.

Down

1.   two points in which the sun crosses the celestial
     equator in its yearly path in the sky.
2.   energy from the Sun humans can see.
3.   the sudden, violent outburst of energy from a star's surface.
5.   a branch of astronomy that deals with the general
     structure and evolution of the universe.
10. is the science that deals with the material
     universe beyond the earth's atmosphere.
11. is a hole, gap, or slit and any other small opening.
     Diameter of the objective of a telescope.

QUIZ

Read the items carefully and shade the box of your choice.

Which is the closest object to Earth?

- ☐ Jupiter
- ☐ Sun

- ☐ Moon
- ☐ Pluto

Finish this sentence correctly: Halley's comet can be seen with the naked eye _____

- ☐ never.
- ☐ every 78 years.

- ☐ 76 times a year.
- ☐ every 76 years.

Which planet is known for its rings?

- ☐ Pluto
- ☐ Venus
- ☐ Saturn
- ☐ Mars

This planet has a giant hurricane-like storm called the Great Red Spot.

- ☐ Uranus
- ☐ Saturn
- ☐ Pluto
- ☐ Jupiter

What are the four terrestrial planets?

- ☑ Mars, Pluto, Jupiter, Ceres
- ☑ Uranus, Neptune, Jupiter Saturn
- ☑ Mercury, Venus, Earth, Mars
- ☑ Mercury, Venus, Earth, Pluto

The outer planets are often called the _____..

- ☑ large comets
- ☑ gas giants
- ☑ asteroid belts
- ☑ galaxy

The order of the Earth Sun and Moon during a lunar eclipse is:

☐ Sun Moon Earth

☐ Earth Sun Moon

☐ Sun Earth Moon

☐ Moon Sun Earth

Some regions of Mars have giant _____.

☐ rivers

☐ lakes

☐ volcanoes

☐ seas

During a lunar eclipse, the moon looks:

- [ ] black
- [ ] reddish orange

- [ ] green
- [ ] pink

What is the last moon phase before the new moon?

- [ ] full moon
- [ ] waning crescent

- [ ] waxing gibbous
- [ ] First quarter

How many planets are in our solar system?

- ■ 9
- ■ 10
- ■ 11
- ■ 8

Pluto was recently reclassified as a _____ planet.

- ■ dwarf
- ■ inner
- ■ outer
- ■ extrasolar

What is the name for the area in between Mars and Jupiter?

- ☐ The Kuiper Belt
- ☐ The Asteroid Belt
- ☐ Ceres
- ☐ Nebula

How many planets in our Solar System have rings?

- ☐ 1
- ☐ 2
- ☐ 3
- ☐ 4

The period of rotation of the Earth:

- [ ] 28 hours
- [ ] 26 hours
- [ ] 24 hours
- [ ] 365 days

What two gases mostly make up Jupiter?

- [ ] oxygen and helium
- [ ] oxygen and nitrogen
- [ ] hydrogen and helium
- [ ] hydrogen and nitrogen

ANSWER

ANSWER

ANSWER

ANSWER

* moon

* every 76 years.

* Saturn

* Jupiter

* Mercury, Venus, Earth, Mars

* gas giants

* Sun Earth Moon

* volcanoes

* reddish orange

* waning crescent

* 8

* dwarf

* The Asteroid Belt

* 4

* 24 hours

* hydrogen and helium

CPSIA information can be obtained
at www.ICGtesting.com
Printed in the USA
LVHW061513240822
PP17497000001BA/3